CANNING AND PRESERVING

A Practical Journal for Life Out Here

Skyhorse Publishing, Inc.

Portions of this book were previously published as *The Complete Book of Home Canning* (978-1-63220-509-4), *Home Canning and Preserving* (978-1-62914-226-5), *Old-Fashioned Jams, Jellies, and Sweet Preserves* (978-1-62914-544-0), and *WECK Small-Batch Preserving* (978-1-5107-3562-0).

10 9 8 7 6 5 4 3 2 1

Library of Congress Cataloging-in-Publication Data is available on file.

Cover design by Melissa Gerber
Cover art provided by Shutterstock.com

ISBN: 978-1-5107-5097-5

Printed in China

This journal belongs to

____

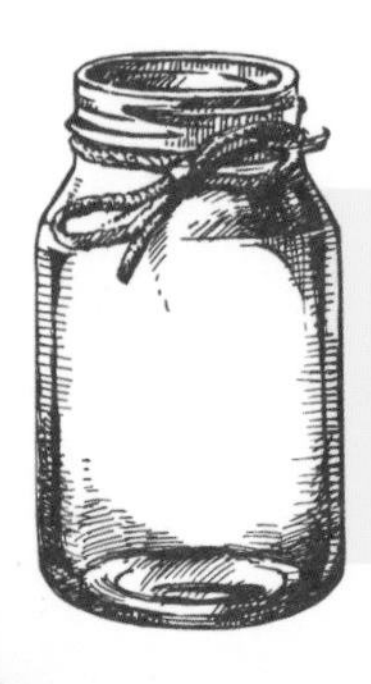

Two things are *essential* for a successfully sealed jar: heat and a clean, sterile jar.

Jams made from a mixture of fruits are usually called *conserves*, especially when they include citrus fruits, nuts, raisins, or coconut.

Jellies that don't set
well can be used to top
ice cream, waffles, or pancakes.

Fruit butter doesn't actually contain butter. Fruit butters are made from *pureed fruit pulp* that is cooked with sugar until thickened to a spreadable consistency.

Sugar serves as a preserving agent, contributes flavor, and aids in gelling. *Cane and beet sugar* are the usual sources of sugar for jelly or jam.

Add new flavors to your jams with ground cinnamon, cloves, nutmeg, ginger, or *red pepper flakes.*

Have a *bumper crop* of tomatoes? Try canning tomato juice, crushed tomatoes, tomato sauce, hot sauce, ketchup, or salsa.

Always process a full load of jars at a time, using *water-filled jars* to fill the empty spaces.

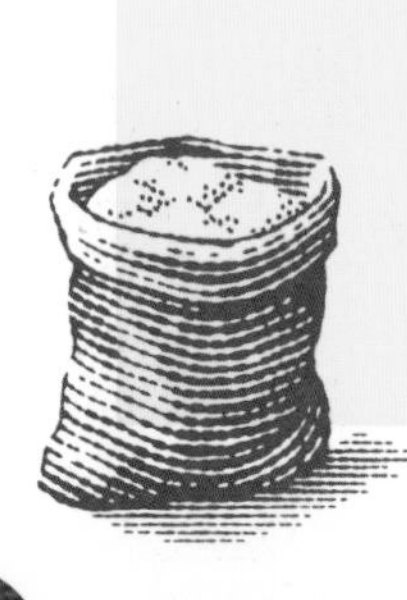

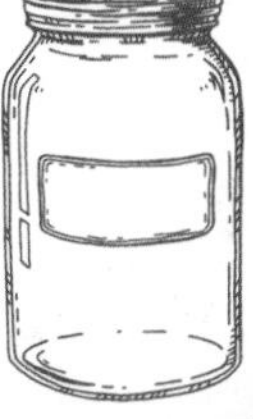

Pectin is the substance in fruits that causes them to jell when combined and cooked with sugar.

Slightly underripe fruit contains more pectin, so use two-thirds ripe fruit with *one-third underripe fruit* for a quicker set and fresher taste.

Don't substitute honey for sugar. *Honey* does not combine well with natural pectin and should not be used in jam or jelly making if a firm set is desired.

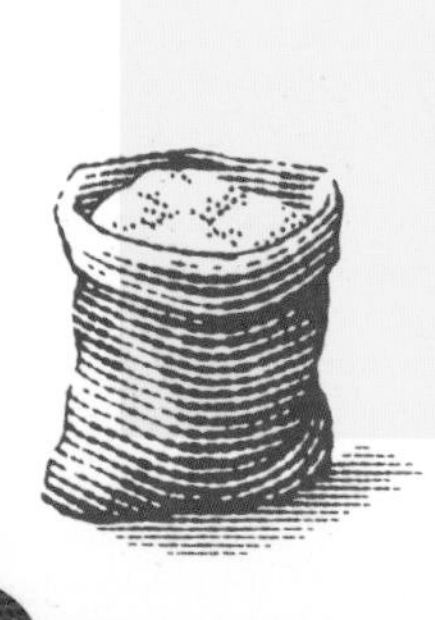

The faster the jelly or jam sets, the more the fruit flavor is retained, so *cook small batches in a large pot.*

Freezer-stored fruit works very well for jam-making if the fruit was picked in *prime condition* and quickly frozen.

Always leave 1/2 inch of headspace (or what is called for in the recipe) between the top of the liquid or food and the rim of the jar. This extra space allows for *expansion of the food* during the sealing process.

When packing jars with fruits or veggies in preparation for the hot water bath, use a sterile metal utensil to *remove any air bubbles* trapped within the contents of the jar to reduce the risk of spoilage.

Always check the rims of the jars and lids to make sure there are *no cracks or chips,* which will interfere with a proper seal.

Leave hot, filled jars where they will not be disturbed for 12 hours. After 12 hours you can *test the seal*. If a jar does not seal after 12 hours, keep it in the refrigerator and use within a couple of weeks.

Store sealed, canned goods in a *dark, dry, and cool place.* It is not recommended to stack your canned goods.

Fruit-based preserves will last about *two months after opening* and vinegar-based preserves are best consumed within six months.

Leave about 1 inch of space between jars in the canning pot so that *water can circulate freely.*

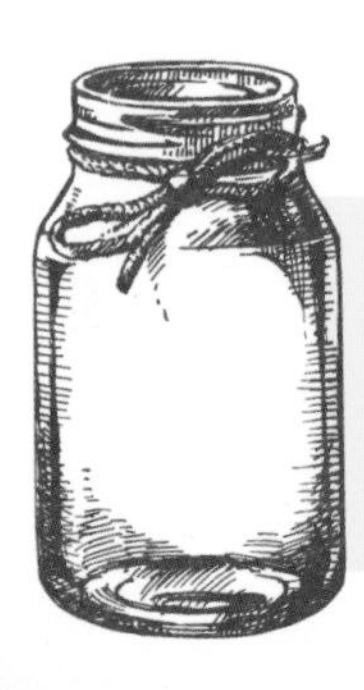

Use paint pens to *label your jars* with the contents and date they were made. The paint is washable, so the jar can be reused.

According to *Guinness World Records,* the largest jam jar is 1,234.14 pounds.

The most *popular item* for canning is tomatoes.

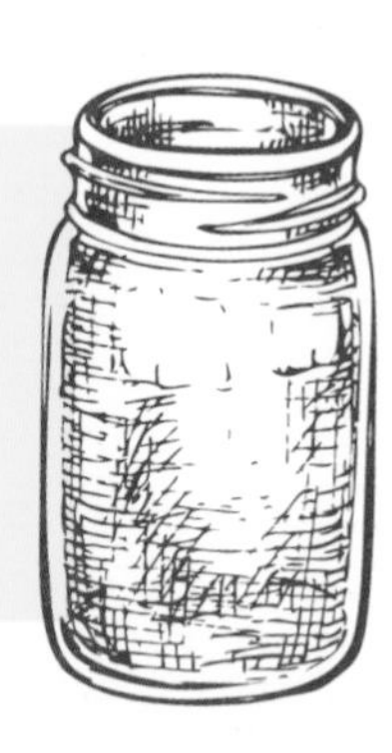

"Putting by" is the colloquial term for preserving food for later use.

To test whether a lid is *properly sealed*, press the middle of the lid with a finger or thumb. If the lid springs up when you release your finger, the lid is unsealed.

If you hold a sealed jar at eye level and look across the lid, the lid should be *concave* (curved down slightly in the center). An improperly sealed lid will be either flat or bulging upward.

When making strawberry jam, do not hull the strawberries until right before you use them.

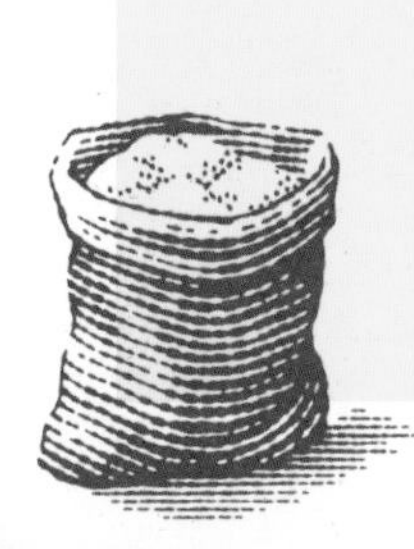

If you are making pickles from cucumbers, be sure to *remove and discard* a 1/16-inch slice from the blossom end of fresh cucumbers. Blossoms may contain an enzyme that causes excessive softening of pickles.

Do not pick raspberries early in the morning, *when the dew is still on them,* because the berries will deteriorate faster.

Red currants and raspberries come into season *at the same time,* and they mix well together in all sorts of preserves.

The larger wild blackberries from the mid-Atlantic states and from the South make a *less seedy jam.*

For peach preserves, select peaches that are *sweet, firm, and very juicy*. Make sure you get freestone peaches; otherwise, you will have a hard time separating the fruit from the pits.

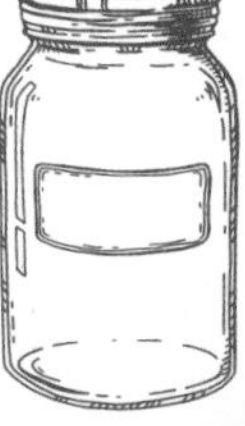

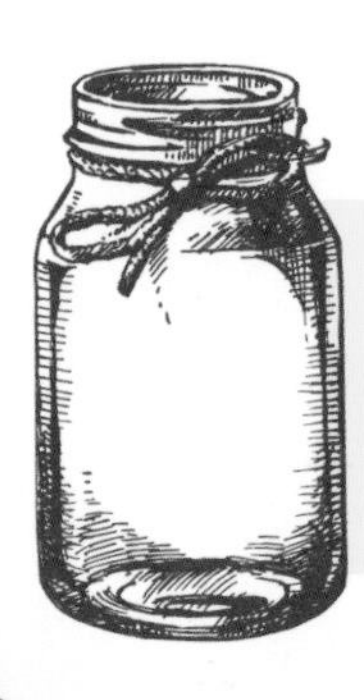

Blueberries are one of the easiest fruits to can, and cooking them seems to *bring out their flavor.* Use firm, ripe blueberries.

Apple butter and applesauce are best made with *tart cooking apples,* such as McIntosh.

Use *a mix of apple varieties* to give your canned apple items more depth of flavor.

In 1858, John Mason invented the first jar that could be used for *canning at home.* It was a glass container with a screw-on thread molded into its top and a lid with a rubber seal.

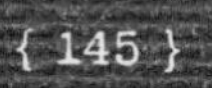

Chutney is a sweet and savory or tangy version of jam. It is made with fruit or vegetables, vinegar, and spices.

Use a cold metal spoon to scoop up some boiling jelly. Tip the spoon. If the jelly runs off the spoon *in a single sheet* (rather than in drops), it is done.

Allow *pickles and preserves* to stand for a few weeks before opening, to allow their full flavor to develop.

Try these *interesting fruit and spice combinations:* apricot and cardamom, cilantro and pineapple, jalapeños and apples, papaya and ginger, blueberries and nutmeg, green tomato and mustard seeds.

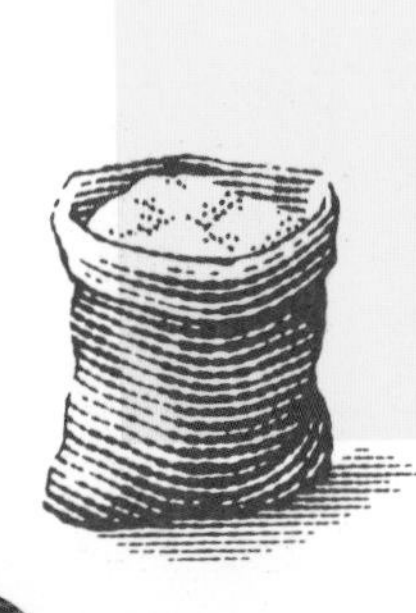

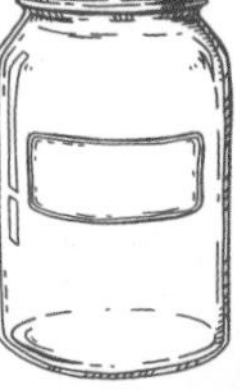

Try these *interesting vegetable and spice combinations:* brussels sprouts and kimchi, leeks and thyme, asparagus and dill seed, beets and cloves, carrots and celery seed, peppers and oregano, zucchini and turmeric.

The best variety of pears for preserving is *Bartlett*, which are widely available in grocery stores.

If you want to make *sugar-free jam,* the best fruits to use (with the lowest carbs and the most natural pectin) are strawberries, blackberries, blueberries, and raspberries.

Include children in the home-canning fun! They can cut soft fruits, measure ingredients, use a cherry pitter to pit cherries, stir ingredients together, and mash fruit with a potato masher.

{ For a unique way to decorate your jars, tie large rounds of *burlap or printed cotton* (gingham prints are a classic) around the top of the jar with twine, ribbon, or bakery string.

Canning lids *should not be reused.* You can reuse the ring that is used to tighten the lids, and the rings can be removed after the jars are properly sealed.

"In seed time learn, in harvest teach, in winter *enjoy.*" —William Blake